AI AGAINST THINKING ERRORS

A Practical Guide to Counter Cognitive Bias

Andre Golard, PhD

Acknowledgements

To Dan Lovallo and Olivier Sibony, whose Case for Behavioral Strategy[1] demonstrates the value of classifying thinking errors. This is used here and was also a major inspiration for my first book.

To Ethan Mollick, whose Co-Intelligence[2] convinced me to become a cyborg.

To my change management mentor Robert Spencer, who explored with me why change is hard[3]. *rip*

---

[1] Lovallo, D. & Sibony, O., 2010. The case for behavioral strategy. McKinsey Quarterly, March issue. https://www.mckinsey.com/business-functions/strategy-and-corporate-finance/our-insights/the-case-for-behavioral-strategy

[2] Mollick, E., 2024. Co-Intelligence: Living and Working with AI, Porfolio.

[3] Golard, A. and Spencer, R., 2013. The Neuroscience of Resistance. Change Management: An International Journal, Volume 12.

Contents

2

3

The Brain's Shortcuts and AI's Solutions

Imagine your brain as a busy chef in a crowded kitchen, constantly looking for shortcuts to serve meals faster. Sometimes these shortcuts create masterpieces. Other times, they lead to half-baked decisions.

Our minds are remarkable but fallible. They take mental shortcuts that can lead us astray:

- We trust the first opinion we hear

- We see patterns where none exist

- We let emotions drive our purchasing decisions

- We convince ourselves we "knew it all along" after the fact

But what if we had a tireless assistant in our mental kitchen – one that never gets emotional, never tires of playing the devil's advocate, and has no personal stake in our decisions?

Enter artificial intelligence.

Throughout history, we've developed tools to combat our thinking errors. Traditional methods like seeking contrary opinions or hiring consultants were valuable but often expensive and time-consuming. More importantly, they carried their own biases and social complications. After all, who enjoys having their cherished ideas challenged?

AI changes this dynamic entirely. It offers instant, unbiased feedback without judgment or emotional baggage. Want to test your hypothesis? Simply ask an AI to disprove it. Looking for blind spots in your thinking? AI can highlight them without fear of hurting your feelings.

This book is your practical guide to better decision-making using AI tools. You won't need to dive into complex technical setups or learn programming. Instead, you'll discover how to ask the right questions—the kind that expose blind spots and illuminate better choices.

A Note on This Book's Creation

While AI tools helped generate comprehensive lists and alternative viewpoints, the narrative you're reading is human crafted. Think of it as a collaboration: AI as the research assistant, human as the storyteller.

On a personal note, this has made writing parts of the book easier, and some much harder given the need to double check and adapt the AI output. I believe the process made the final product much better.

Two Paths to Better Thinking

You can address thinking errors in two ways:

1. Tackle current decisions with immediate AI feedback

2. Establish ongoing processes for recurring decisions

This book focuses on the first approach—immediate solutions for pressing decisions. While powerful tools exist for systematic decision screening, they evolve too rapidly for print. Besides, most tools are domain-specific rather than universal.

The book starts with a review of basic thinking processes. The second part investigates how AI can be incorporated into these processes.

The focus is at first on the individual. The same principles apply to organizations.

A Word of Caution Experiment with different prompts and AI tools. Some will provide brilliant insights; others might "hallucinate" or generate unreliable responses. The key is learning which approaches work best for your specific needs.

Most importantly: While AI helps us think faster, the most crucial decisions still deserve time and reflection. The difference now is that you can use these tools to make that reflection more productive and insightful than ever before.

Welcome to the future of decision-making—where artificial intelligence helps us be more genuinely human.

PART 1: UNDERSTANDING COGNITIVE BIASES

Common cognitive biases

Whenever I tell people I am writing a book on thinking errors, many answer "I'm an expert at that!" Here are a few examples, along with standard ways to deal with them.

The Stories We Tell Ourselves

The human mind is a masterful storyteller, but sometimes it's a terrible fact-checker.

I remember the day I first realized how deeply I could deceive myself. It was a crisp December morning in Seattle, and I was absolutely convinced that my favorite football team, the Seahawks, would make it to the Superbowl. Every piece of news, every statistic I encountered, I filtered through the lens of my predetermined belief. When sports analysts suggested vulnerabilities in the team's lineup, I dismissed them. When friends raised doubts, I countered with cherry-picked statistics.

We lost in the first round of the playoffs. Spectacularly.

That painful moment was my introduction to confirmation bias – a cognitive error that transforms our brains into echo chambers of our own beliefs. Confirmation bias is the tendency to search for, interpret, favor, and recall information in a way that confirms or supports one's prior beliefs or values. It's a mental shortcut that makes us comfortable but dangerously blind.

Where Do You See Yourself?

Close your eyes. Think of a strongly held belief. Maybe it's about:

- Politics
- Your partner
- Your career potential
- A family member
- A personal capability

Now ask yourself: What evidence are you selecting? What are you ignoring?

Confirmation bias is the mind's remarkable ability to see what it wants to see. It's our brain's default setting—a cognitive shortcut that transforms us into lawyers defending a predetermined verdict instead of investigators seeking the truth.

Imagine your brain as a selective librarian, carefully curating evidence that supports your existing beliefs while quietly discarding anything that might challenge them. It's not malicious. It's survival. Our brains evolved to create quick, coherent narratives, not necessarily accurate ones.

Real-World Confirmation Bias Landscapes

This cognitive error shows up everywhere:

- Investors believing in a failing stock despite clear market signals

- Political supporters consuming only media that reinforces their worldview

- Medical patients ignoring symptoms that don't fit their preferred diagnosis

- Managers who hire candidates most like themselves

Your Confirmation Bias Workout

Breaking free requires radical intellectual honesty:

- *Actively seek contradictory information*
    - Read sources that challenge your beliefs
    - Ask, "What if the opposite were true?"

- *Practice intellectual humility*
    - Admit "I might be wrong"
    - Reward yourself for changing your mind

- *Create a "Devil's Advocate" file*
    - Write down arguments against your current beliefs
    - Review them quarterly

- *Diversify your information sources*
    - Follow people with different perspectives
    - Listen more than you speak

- *Ask transformative questions*
    - "What evidence would change my mind?"

o   "How might someone with a different
        background see this?"

Personal Reflection

Take out a piece of paper. Draw a line down the middle.

- Left side: Write a belief you're certain about
- Right side: List three pieces of evidence you might be overlooking

Remember: The most dangerous phrase in the language is "We've always done it this way."

As the brilliant cognitive scientist Daniel Kahneman observed: "What you see is all there is[4]" – but it doesn't have to be.

The Dangerous Comfort of Consensus

Conformity is a poison that tastes like belonging.

The conference room felt electric with unspoken tension. Our startup was about to launch a ˜evolutionary product, and everyone – everyone except me – seemed convinced it was perfect. When I raised my hand to suggest a potential flaw in our design, the room went cold. Looks were exchanged. Subtle eye rolls. Barely concealed sighs.

---

[4] Kahneman, D., 2011. Thinking Fast and Slow.

I shut up. We launched. The product failed spectacularly.

Recognize Yourself?

Take a moment. Think about a time you:

- Stayed silent in a meeting
- Went along with a group decision you secretly doubted
- Suppressed your true opinion to avoid conflict
- Felt uncomfortable challenging the status quo

That uncomfortable moment? That's groupthink in action.

Groupthink is the psychological phenomenon where the desire for harmony and conformity overwhelms critical thinking. It's a collective delusion where belonging trumps truth, and agreement becomes more important than reality.

Imagine your brain as a social chameleon, constantly negotiating between individual thought and group survival. Groupthink is what happens when the chameleon decides it's safer to blend in completely—even if blending means walking off a cliff.

Where Groupthink Lurks

This silent killer appears everywhere:

- The Space Shuttle Challenger disaster, where engineers' warnings were suppressed.

- Corporate scandals like Enron, where questioning leadership was career suicide.

- Political echo chambers where dissent is viewed as betrayal.

- Military strategies developed without challenging core assumptions, such as the Bay of the Pigs fiasco.

Your Groupthink Resistance Training

Breaking free requires courage and systematic approaches:

- *Create Psychological Safety*
  - Ensure team members can speak without fear of retribution
  - Reward people for speaking up
  - Make "I'm not sure" a valid response

- *Encourage devil's advocates*
  - Explicitly assign someone to challenge group assumptions

- *Diversify Your Circles*
  - Seek perspectives different from your own
  - Listen without defending
  - Practice genuine curiosity

- *Develop a Personal "Doubt Protocol"*
  - When everyone agrees, ask: "What are we missing?"

- o   Challenge yourself to find at least one alternative perspective

- Build Your Dissent Muscles
  - o   Start small: Speak up in low-stakes situations
  - o   Celebrate your courage, not the outcome

- Use anonymous feedback mechanisms
  - o   Allow candid input without social pressure

- Practice Constructive Disagreement
  - o   Use "and" instead of "but" – learn the Yes, And of improv theater
  - o   Focus on solving, not winning[5]

---

[5] Picture this: A struggling mine in Wisconsin. Two groups locked in a battle of wills – the local team versus headquarters. Everyone's shouting their opinion, nobody's listening, and important decisions are stuck in the mud. Sound familiar?

Enter Roger Martin, a young consultant who would later become a business school dean. Instead of letting the chaos continue, he pulled a genius move.

"Let's try something different," he said, cutting through the noise. "Instead of arguing about who's right, let's look at each option and ask: What would have to be true for this to be the best choice?"

The result? Pure magic.

Suddenly, people stopped defending their turf and started exploring possibilities. The conversation shifted from "Here's why you're wrong" to "Here's what we'd need to see to prove you right." It was like flipping a switch from combat to collaboration.

Think about it: Instead of asking "Should we close the mine?" (which gets everyone's defenses up), they asked "What conditions would make closing the mine the smart choice?" This subtle shift changed everything.

      o   Separate the idea from the person

Personal Reflection

Draw a line down a page.

- Left side: Describe a recent group decision

- Right side: List potential blind spots or alternative approaches

Remember: Consensus is not truth. Comfort is not correctness.

As Irving Janis, who first described groupthink, warned: "The more amiable the group, the more likely it is to engage in faulty decision-making[6]."

The Price of Holding On

A single dollar lost hurts more than a dollar gained feels good.

I stared at the stock market app, my finger hovering over the "sell" button. The tech stock I'd held for years was tanking, down 40% from its peak. Every rational indicator suggested

---

Next time you're stuck in a meeting where everyone's talking but nobody's listening, try this trick. Don't ask who's right – ask what would need to be true for each option to be the winner. Watch how quickly the mood changes from fighting to finding solutions.
Sometimes the best way to win an argument is to stop arguing altogether.

[6] Janis, I., 1972. Victims of Groupthink. Houghton Mifflin.

cutting my losses, but something deep inside me screamed to hold on – to avoid admitting defeat.

Three years later, that stock was worth less than half of what it had been when I first hesitated.

Do You Recognize This Feeling?

Close your eyes. Think about something you're holding onto:

- A relationship that no longer serves you

- A job that's draining your spirit

- An outdated belief

- An old possession with more emotional weight than practical value

- A past mistake you can't forgive yourself for

That hesitation you feel? That's loss aversion whispering in your ear.

Loss aversion is our psychological tendency to feel the pain of losing something about twice as intensely as the pleasure of gaining something of equal value. It's an evolutionary survival mechanism gone rogue – protecting us from potential threats by making us terrified of letting go.

Imagine your brain as a hyper-cautious museum curator, desperately protecting every exhibit, even the ones gathering dust and preventing new, vibrant displays from taking their place.

Where Loss Aversion Hides

This silent saboteur appears in countless scenarios:

- Investors holding onto failing stocks

- People staying in unfulfilling relationships

- Employees clinging to outdated skills

- Consumers keeping unused subscriptions

- Individuals refusing to change destructive habits

Your Loss Aversion Liberation Plan

Breaking free requires strategic, compassionate action:

- *Quantify Opportunity Costs*
  - Calculate the real price of holding on
  - Ask: "What am I preventing by keeping this?"

- *Practice Emotional Detachment*
  - View decisions as experiments, not verdicts
  - Separate your worth from your possessions or past choices

- *Create Letting-Go Rituals*
  - Develop personal ceremonies for releasing[7]
  - Celebrate the courage of transformation

---

[7] Check out my friend Danny Dover's The Minimalist Mindset, 2017.

- *Reframe Your Narrative*
    - Focus on potential gains, not potential losses
    - Ask: "What becomes possible if I let this go?[8]"
- *Build a Support System*
    - Share your letting-go journey
    - Seek perspectives from those not emotionally involved

Personal Reflection

Take a piece of paper. Draw a line down the middle.

- Left side: Write something you're afraid to lose
- Right side: List three potential opportunities that could emerge from letting go

Remember: The empty hand is the one most ready to receive.

As Nobel laureate Daniel Kahneman wisely observed: "Losses loom larger than gains[9]" – but they don't have to define you.

---

[8] See Loss Aversion is Real, Allison, T. & Golard, A., 2012. Financial Advisor https://www.fa-mag.com/news/loss-aversion-is-real-16449.html
[9] Kahneman, D. & Tversky, A., 1979. Prospect Theory: An Analysis of Decision under Risk. Econometrica. 47 (4): 263–291.

The Tunnel Vision of Decision-Making

Sometimes, the biggest obstacle to solving a problem is how we look at it.

The renovation was supposed to be simple. My wife and I wanted to update our kitchen, and I fixated on a single contractor's quote. Fifteen thousand dollars seemed astronomical. I spent weeks agonizing, comparing minute details, searching for the cheapest option –completely blind to the bigger picture of value, quality, and long-term satisfaction.

When we finally completed the project with a different contractor, we realized we'd actually saved money by choosing quality over the lowest price. My narrow focus had nearly cost us a much better outcome.

Sound Familiar?

Take a moment. Close your eyes. Think about a recent decision that felt overwhelming. Maybe it was:

- Choosing a new job

- Buying a car

- Selecting a college

- Planning a major life change

Chances are, you've experienced narrow framing without even realizing it. That gut-wrenching moment when you feel stuck, seeing only one path, one option, one perspective—that's narrow framing in action.

Narrow framing is the cognitive trap of making decisions by focusing too closely on a single aspect while overlooking the broader context. It's like trying to understand an entire landscape by peering through a keyhole – you miss the vast, crucial details that provide true perspective.

Where Do You See Yourself?

This mental limitation manifests in numerous ways:

- Investors choosing individual stocks without considering overall portfolio strategy

- Businesses cutting costs in one department without analyzing company-wide impact

- People selecting job offers based solely on salary, ignoring work-life balance and growth potential

- Individuals making health decisions by treating symptoms rather than understanding root causes

Your Narrow Framing Workout

Breaking free requires asking yourself powerful questions:

- *Zoom out regularly*: "What am I not seeing?"

- *Challenge your first instinct*: "If a friend were in my situation, what advice would I give them?"

- *Imagine alternative scenarios*: "What if the opposite were true?"

- *Seek outside perspectives*: "Who might see this situation differently?"

- *Connect the dots*: "How might this decision impact other areas of my life?"

A Personal Reflection

Take out a piece of paper. Draw a line down the middle. On the left, write a recent decision that felt constraining. On the right, list three broader perspectives you might have missed.

Your brain is a powerful tool, but it can also be your greatest limitation. Narrow framing isn't a failure – it's an opportunity to grow, to learn, to see the world in technicolor instead of black and white.

As management expert Peter Drucker wisely observed: "The most serious mistakes are not being made as a result of wrong answers. The most serious mistakes are being made as a result of wrong questions[10]."

---

[10] Drucker, P., 2010. Men, Ideas, & Politics, Harvard Business Review Press.

Types of cognitive biases

Why classify biases

In McKinsey's "The Case for Behavioral Strategy[11]", Dan Lovallo and Olivier Sibony spelled out four classes of thinking errors based on the business processes associated with them. This was the inspiration for my first book, "A Field Guide to Thinking Errors". Having classification allows us to look for common solutions. For example, knowing that loss aversion is linked to the activation of our fear centers means that we can counter it using practices designed to reduce fear in other domains.

Rather than focusing on business processes I chose to investigate errors individuals face in everyday life. I am a neuroscientist, so my classification is based on brain processes.

A brief introduction to our brain

Imagine your brain as an archaeological site, with layers of evolutionary history buried within its intricate landscape. Each layer represents a different stage of development, a testament to the remarkable journey of survival and adaptation that has shaped our species.

---

[11] Lovallo, D. & Sibony, O., 2010. The case for behavioral strategy. McKinsey Quarterly, March issue. https://www.mckinsey.com/business-functions/strategy-and-corporate-finance/our-insights/the-case-for-behavioral-strategy

The triune brain model—proposed by neuroscientist Paul D. MacLean in the 1960s[12,13] – is not just a theory. It's a profound narrative of how we became human, a story written in the very architecture of our neural networks.

Picture three distinct but interconnected brain regions, each representing a different era of evolutionary complexity:

- The reptilian complex: our most ancient survival system, responsible for instinctual behaviors

- The limbic system: the seat of our emotional intelligence, responsible for the motivation and emotion involved in feeding, reproductive behavior, and parental behavior

- The neocortex: the pinnacle of human cognitive capability, conferring the ability for language, abstraction, planning, and perception

Yet, within this model, a fourth region emerges as a marvel of evolutionary sophistication: the prefrontal cortex. Located at the very front of the brain, this region represents the most advanced stage of our neural development – the crown jewel of human consciousness.

The prefrontal cortex is our brain's executive center. It's responsible for:

---

[12] Further developed in MacLean, P.D., The Triune Brain in Evolution, 1990.
[13] Please note that the concept is no longer regarded as valid in terms of evolution but remains a useful, basic way to classify brain areas.

- Complex decision-making

- Personality expression

- Moderating social behavior

- Planning future actions

- Impulse control

- Emotional regulation

Unlike the older brain regions that operate on instinct and immediate response, the prefrontal cortex introduces a revolutionary capacity: the ability to pause, reflect, and choose. It's the neural equivalent of a wise mediator, constantly negotiating between our primal impulses and our higher-order thinking.

When you resist an immediate temptation, carefully consider a difficult decision, or empathize with someone else's perspective, your prefrontal cortex is hard at work. It's the part of your brain that allows you to be more than just a collection of reflexes and emotions – it's what makes you truly human.

These are not just abstract regions. They are living, breathing systems that continue to interact, compete, and collaborate within the theater of our consciousness. The reptilian brain stands guard, the limbic brain feels, the neocortex thinks, and the prefrontal cortex reflects and chooses – sometimes in harmony, sometimes in conflict.

When you feel an inexplicable surge of fear, that's your reptilian brain. When you're moved to tears by a piece of music, your limbic system is at work. When you solve a complex mathematical equation, your neocortex takes center stage. And when you take a deep breath and choose a measured response instead of an angry outburst, your prefrontal cortex is guiding you.

This is not just neuroscience. This is the story of how we became who we are.

So, how do you apply this to countering our thinking errors?

Let's look back at the four errors introduced in the previous section.

The stories we tell ourselves describes the confirmation bias.

It is based on the way our *cortex* stores and recalls memories.

Other errors in this category include:

- Anchoring – taking the first value given and using it for future deliberations
- The availability bias - our tendency to overestimate the likelihood or importance of events that are easier to recall from memory
- Confusing correlation with causation
- False memories

- The halo effect - the tendency to let one positive trait or impression of a person, brand, or thing influence our overall evaluation of their other characteristics
- The hindsight bias – "I knew it!"
- The omission bias – the tendency to judge harmful actions as worse than equally harmful omissions or inactions
- Overconfidence
- Seeing patterns when there are none

Making Better Decisions: A Guide to Outsmarting Your Brain

We all make snap decisions – it's how our brains are wired. Like Daniel Kahneman explains in his bestseller[14], we have two ways of thinking: quick and automatic, or slow and deliberate. The trouble is, those quick decisions often lead us astray.

So how can we make better choices? First, take a breath. Give yourself permission to slow down. Try viewing your situation through different lenses – maybe you're missing something obvious. If you think A is causing B, pause and ask: could there be a hidden factor C that's behind both?

Here's a practical tip: keep a decision journal. Write down what you think will happen before making big choices. When you look back later, you might be surprised (or even embarrassed) by how often you were wrong. But that's okay – it's how we learn.

---

[14] Kahneman, D., 2011 Thinking Fast and Slow. Penguin.

Context is everything. Sometimes what looks like a terrible decision made perfect sense given what people knew at the time. That's why Monday morning quarterbacks – whether in sports or on TV – often miss the point. They're great at spinning compelling stories, but their predictions usually aren't any better than yours or mine.

Speaking of stories – our brains love them. They're like mental shortcuts, helping us make sense of the world. But these shortcuts can trick us. Did you hear about a shark attack on the news? Suddenly every beach seems dangerous, even though the actual risks haven't changed one bit.

Challenge yourself. Seek out voices that disagree with you. Ask friends for their honest opinions but be careful how you ask. "What do you think about this?" will get you better insights than "Don't you agree that...?"

Question your habits. Why do you buy that particular brand? Take that specific route to work? Sometimes we do things simply because we've always done them that way. Mix things up occasionally – you might discover better alternatives.

Watch out for manipulation. That "original price" crossed out on a sales tag? That carefully curated set of options? They're designed to nudge you toward certain choices. Step back and ask what's missing from the picture.

The good news is that we can train ourselves to make better quick decisions. Expert firefighters do this through practice scenarios – they learn to read situations instantly. But for the big choices in life – career moves, relationships, major purchases – there's no shame in taking it slow.

Remember: It's not about second-guessing every decision. It's about knowing when to shift from autopilot to manual control. Your brain's shortcuts usually serve you well – they're just not perfect. The trick is knowing when to override them.

The most valuable lesson? We know less than we think we do. And that's perfectly fine – as long as we remember it when it matters most.

The Dangerous Comfort of Consensus is about groupthink.

It is about how we interact with each other.

Other errors in this category include:

- Believing pundits – The tendency to give undue weight to the opinions of media personalities and commentators who speak with authority, despite their frequent lack of relevant expertise or accountability for incorrect predictions.
- Believing second-hand information.
- The fundamental attribution error – if things work out it is because I'm a genius. If they don't it is due to external factors. We tend to overemphasize

personality-based explanations for others' behaviors while underestimating the influence of situational and environmental factors.

- Tribalism - the tendency to strongly identify with and show loyalty to one's own group while viewing other groups with suspicion or hostility.

Breaking Free from the Echo Chamber: A Guide to Better Social Thinking

We humans are tribal creatures. It's in our DNA – a leftover from ancient times when sticking with our group meant survival. But in today's world, this instinct can lead us astray. Let me share some insights on breaking free from social blind spots while keeping your relationships intact.

Think about your "tribes" for a moment. Maybe it's your political party, your neighborhood group, or your online community. These groups shape how we think more than we realize. And thanks to social media, it's easier than ever to surround ourselves with people who think exactly like we do. Comfortable? Yes. Healthy? Not always.

Here's the thing: disagreeing with your group doesn't have to mean exile. Unlike our cave-dwelling ancestors, we won't starve if we voice a different opinion. In fact, you might find it liberating. Try dipping your toes in new waters – join a hobby group, volunteer at a local shelter, or take up community gardening. These new connections can broaden your perspective without threatening your existing relationships.

When conflicts arise (and they will), remember that nobody likes being told they're wrong. Instead of arguing, try this: Listen first. Really listen. Then reflect back what you've heard – "So what you're saying is..." This simple act can transform a potential argument into a conversation. You might be surprised how often finding common ground leads to better solutions for everyone.

Want to test your information diet? Ask yourself:

- Where is this information coming from?
- Who benefits if I believe this?
- Am I seeing the whole picture, or just a carefully selected piece?
- What would someone from outside my usual circle think about this?

Remember that old saying "divide and conquer"? Some people profit from keeping us at each other's throats. Don't let them win.

For those working in teams, diversity isn't just a buzzword – it's your secret weapon against groupthink. Mix up your team. Encourage the quiet ones to speak up. Try breaking into smaller groups before tackling big decisions. And here's a powerful trick: imagine your project has already failed miserably. What went wrong? This "pre-mortem" can spot potential problems before they happen[15].

The world isn't black and white, and neither are most solutions. Next time you're faced with a yes/no choice, pause and ask: "What other options am I missing?"

Breaking free from social biases isn't about abandoning your groups or beliefs. It's about growing beyond them while keeping what matters. After all, the best tribes are those that help us become better versions of ourselves.

What new perspective might you gain by stepping outside your comfort zone today?

The Price of Holding On is about loss aversion.

This happens when our emotions dominate.

Other errors in the category include:

- Discounting future rewards – the tendency to prefer smaller, immediate benefits over larger, future ones, even when waiting would be more rational or beneficial.
- Gambling – We sometimes take big risks, ignoring the odds. For some of us, it is a recurring problem.

---

[15] Klein, G., 2007. Performing a project premortem. *Harvard Business Review, September issue*.

- "I know best" – the tendency to overestimate one's own judgment and knowledge while dismissing others' perspectives or expertise.
- Inadequate rewards – Ask a friend to drive you to the airport, and he/she might. Offer them $5 to drive you, and the friend more likely will remind you of the existence of taxis and buses.
- Overvaluing what we have – also known as the endowment effect is our tendency to place higher value on things we own compared to identical items we don't possess
- Procrastination – I will find a description later.
- The status quo bias – our irrational preference for keeping things the same, even when making a change would be better for us.
- Staying the course – similar to the status quo bias but implies an action rather than inaction.

Mastering Your Emotions: A Practical Guide to Smarter Decisions

Ever notice how different your choices look when you're stressed versus when you're calm? There's a reason for that. Our brains have an emotional autopilot that's lightning-quick but not always wise. Here's how to take back the controls when it matters most.

Start by hitting the pause button. I know, easier said than done. But here's the secret: your best thinking happens when you're rested, well-fed, and not stressed. It's like trying to have a serious conversation at a rock concert versus over a quiet cup of coffee – which one do you think will go better?

For decisions you make often, create a game plan ahead of time. Think of it like meal prepping for your brain. When emotions are running high in group settings, shift the focus to facts and analysis. Make it less about "who's right" and more about "what's right." And always leave the door open for someone to say, "Hey, maybe we should look at this again."

Before making big moves, ask yourself: How much do I really know about this? Have I done this before? Did I learn anything useful from past attempts? Sometimes we feel like experts just because something feels familiar.

Fear is a tricky customer. It's like having multiple tabs open in your brain – each worry adds up until your mental computer starts lagging. Politicians know this; they love hitting those fear buttons because it shuts down our rational thinking.

Try this mental flip: Instead of obsessing over what you might lose, ask yourself how hard you'd work to gain that same thing. Scared to switch jobs? Imagine if you were unemployed – how excited would you be about this opportunity?

Starting small works wonders. Can't face a complete career change? Take a night class. Nervous about investing? Start with a tiny amount. Each small win builds confidence for bigger steps.

Here's a powerful trick: when weighing options, don't just think about them – feel them out. Close your eyes and imagine living with each choice. Your gut often knows things your mind hasn't figured out yet.

Remember, what's precious to you might be worthless to someone else, and vice versa. That's not wrong – it's just human. The key is recognizing when your values are helping you make good choices and when they're holding you back.

Want to build emotional resilience? Practice stepping out of your comfort zone in small ways. Chat with someone from a different background. Sell a tiny bit of that stock you've been clutching. Each small step makes the next one easier.

The goal isn't to become an emotionless robot. It's about finding that sweet spot where your feelings inform your decisions rather than hijack them. After all, the best choices often come from combining head and heart – just make sure they're both getting a fair say.

Ready to take that first small step toward better decisions?

The Tunnel Vision of Decision-Making is about narrow framing.

It is due to the limited capacity of our prefrontal cortex, which can lead us to oversimplify.

Other errors in this category include:

- Bad approximations
- False alternatives
- Multi-tasking
- Overthinking

Your Brain's Command Center: A User's Guide to Better Decisions

Think of your brain's prefrontal cortex as your personal CEO. It's brilliant at complex decisions, but like any executive, it has its limits. Let me show you how to work with these limits rather than against them.

First, the basics: Your mental CEO can juggle about seven things at once – think of it like keeping seven balls in the air. This ability peaks in your early twenties (sorry, folks!), and like a smartphone battery, it needs regular recharging. That means food, rest, and occasional timeouts.

Here's where most of us trip up. Either we oversimplify ("Should I do this or not?") when we need more options, or we get paralyzed by too many choices ("Which of these 47 phone plans is best?"). Neither extreme serves us well.

So, what's the fix? Let's start with some practical tips:

Do one thing at a time. Your brain isn't built for multitasking, no matter what your resume claims. When you need to make an important decision, give it your full attention. Think of it like driving – you wouldn't text and navigate a busy intersection, right?

Is a solution popping into your head too quickly? Red flag! Ask yourself: "What if my first idea wasn't an option?" Sometimes our best decisions come from our second or third thoughts.

Feeling overwhelmed? Break it down:

- Get it out of your head and onto paper (or screen)
- Tackle it in chunks, like eating an elephant one bite at a time
- Compare options in pairs – it's easier to choose between A and B than A through Z
- Draw a simple decision tree, eliminating obvious non-starters first
- Step back and look at the bigger picture

For everyday decisions, create simple rules. Which cereal to buy? "I'll pick the third one on sale from the left." Done. Save your mental energy for bigger things.

Want to boost your brain's capacity? Be wary of miracle solutions promising to "train your brain." Most only make you better at specific tasks – like getting really good at Sudoku but still struggling with real-world problems. Instead, try working through real-life scenarios and case studies. Think of it as a flight simulator for life decisions.

And remember: if you're stressed, emotional, or trying to avoid losses, your mental CEO might be temporarily offline.

That's when you need to take a break, have a snack, and come back fresh.

The goal isn't to become a decision-making machine – it's to work with your brain's natural abilities while respecting its limits. After all, even CEOs need lunch breaks.

Want to start improving your decisions today? Begin by decluttering your environment. A tidy space leads to tidier thinking. Sometimes the simplest changes make the biggest difference.

PART 2: AI FOR COGNITIVE BIAS MITIGATION

Introduction: AI as a Mirror for the Mind

We all like to think we're rational beings, making careful decisions based on facts and logic. But here's an uncomfortable truth: our brains are magnificent machines that sometimes play tricks on us. They take shortcuts, jump to conclusions, and occasionally lead us astray – all in the name of efficiency.

Enter artificial intelligence, our newest intellectual companion. AI systems, in their relentless logic and pattern-matching capabilities, have become unexpected mirrors, reflecting back our own thinking quirks and biases. By understanding how these machines think – and sometimes fail – we can better understand our own mental blind spots.

Think about confirmation bias, our tendency to seek out information that confirms what we already believe. When AI systems were first trained on human-generated data, they started picking up our biases too. Imagine a resume-screening AI that favored candidates from certain universities simply because that's what past hiring managers had done. The machine wasn't malicious; it was simply learning from our own flawed patterns.

But here's where it gets interesting. Once we spotted these biases in AI systems, we had to figure out how to correct them. And in doing so, we learned valuable lessons about our own thinking. If we can teach machines to check their assumptions and consider multiple perspectives, couldn't we do the same for ourselves?

Consider how AI approaches a problem: it systematically evaluates multiple possibilities, assigns probabilities, and updates its conclusions based on new evidence. No emotional attachments, no ego to defend, no deeply held beliefs to protect. While we wouldn't want to think exactly like machines (our emotions and intuitions are valuable too!), we can learn from their methodical approach.

The next time you're faced with a difficult decision, try thinking like an AI:

- What data am I working with?

- What assumptions am I making?

- What alternative explanations haven't I considered?

- How would I update my thinking if I received new information?

The rise of AI isn't just about creating smarter machines – *it's about becoming more aware of how we think and reason.* By understanding both the strengths and limitations of artificial minds, we can better appreciate and improve our own thinking processes.

In the previous chapters we explored cognitive biases and how we typically deal with them. In the chapters that follow, we'll explore practical strategies for clearer thinking that involve AI. Together, we'll discover how the age of AI can help us become not just smarter, but wiser decision-makers.

Because in the end, the greatest gift AI might give us isn't the ability to automate tasks or process vast amounts of data – it's the mirror it holds up to our own minds, showing us both our brilliance and our blind spots.

Please keep in mind that different people will have different comfort levels with using AI. When you see an AI prompt in the following chapters, ask yourself "is this a question I could ask my team"? Then ask AI anyway and see if the two answers differ. This bridges the gap between using AI and better management and leadership.

Let AI Be Your Mental Detective: A Guide to Clearer Thinking

We have explored traditional ways to counter individual types of errors. AI can be your ally in spotting and outsmarting these mental traps. Here's how to leverage digital tools to sharpen your thinking:

For Confirmation Bias: Let AI be your devil's advocate. Ask it to present counterarguments or find information that challenges your current beliefs. Ask AI to play fact-checker. "Show me the actual data behind this trend I think I'm seeing" or "What evidence might prove me wrong?" Think of it as training wheels for seeing both sides of any story.

For Group Think: Before team meetings, gather diverse perspectives from AI to ensure you're considering views outside your usual echo chamber.

For Emotional Blind Spots: Use AI as your calm, rational friend. When emotions are running high, ask it to list pros and cons, or to analyze a situation from multiple angles. It's like having a level-headed advisor who never gets caught up in the moment.

For narrow framing: Use AI to break down complex decisions into smaller, manageable chunks. Have it create lists of alternatives "What are five completely different approaches to this situation?" or "Show me how others have successfully changed similar situations."

Remember: AI isn't here to think for you – it's here to help you think better. Use it as a mirror to reflect your biases and a spotlight to illuminate blind spots. The goal isn't perfect thinking (that doesn't exist), but rather more balanced, thorough consideration of your choices.

Let's look at each type of error in more detail. **Consider that the tools and prompts for errors of the same type are likely to be similar**. See Classifying Thinking Errors[16]. Adapt as you see fit. Tools and prompts may also apply to more than one class of errors and thus may be mentioned more than once. They are included for completeness. Skip as you see fit.

Considerations for the tools

- Tools are constantly evolving. What is available at press time will include tools not covered here.
- Tools have a learning curve. Some may be best suited to an organization rather than an individual. See what makes sense for you.
- Some tools will have a paywall for full features. As an individual you have many options including free ones to explore.
- Tools that are most effective for you will depend on your unique situation. Try them, select a handful, and learn to use them well.

---

[16] pp17-18

- Not all tools listed are AI-powered. If a spreadsheet plugin is useful to avoid an error it will be included.

So, let's jump in and explore tools and prompts to help you with different types of thinking errors. Each section follows the same format: first list tools, then list prompts.

Confirmation Bias: Breaking Your Brain's Echo Chamber

Let's face it – we all love being right. So much that we often cherry-pick facts that support what we already believe. But in today's complex world, that's a luxury we can't afford. Be aware that some of the tools listed (such as Power BI) have a significant learning curve. These may be best for the enterprise rather than the individual.

Here's how to use AI as your personal bias-buster:

Tools

- *Tools to change the angle of vision*

  o AI-powered Search Engines: Tools such as **PerplexityAI** can provide diverse perspectives by surfacing articles and viewpoints that challenge your initial search query. Some search engines are starting to incorporate features that highlight different viewpoints.

  o **Google Scholar/ Semantic Scholar**: Use these to access peer-reviewed research and diverse academic sources.

- Expanding your information diet

    - **AllSides**: Provides news coverage from multiple political perspectives.

    - **Feedly**: Curates content from a wide range of sources to reduce echo chambers.

    - **Ground News**: Compares news coverage and highlights potential biases or blind spots.

- *Argument Mapping Software*

    - Tools like **Rationale** or **MindMup** allow you to visually map out arguments and counterarguments, making it easier to identify weaknesses in your own reasoning and consider opposing viewpoints.

- *Fact-Checking Bots and Websites*

    - Sites like **Snopes, PolitiFact**, Poynter's **MediaWise,** and **FactCheck.org** can help you verify information and avoid relying on biased or inaccurate sources.

    - **Truth Goggles**: Uses neuro-linguistic programing (NLP) to analyze text for biases or inaccuracies.

- *AI-powered Debate Platforms*
  - **Kialo**: A debate and discussion platform for exploring opposing views.
  - **Polis**: AI-powered platforms for gathering and analyzing public opinion on contentious issues.

  While still emerging, these platforms could be used to simulate debates with AI opponents who are trained to argue from different perspectives.

Prompts

These prompts can be used with AI tools or as self-reflection exercises:

General Prompts
- "What evidence contradicts my current belief on [topic]?"
- "What are the strongest arguments against my position on [topic]?"
- "If I were trying to argue the opposite side of this issue, what points would I emphasize?"
- "What are alternative explanations for the evidence I'm considering?"
- "What assumptions am I making, and are they valid?"
- "What are the potential downsides of being wrong about this?"

Specific Prompts

- **For Evaluating Information Sources**

    - "Analyze this article for potential biases. What perspectives are included, and what perspectives are missing?"

    - "Find sources that present a different viewpoint on the claims made in this article."

    - "What is the source's reputation for accuracy and objectivity?"

- **For Decision-Making**

    - "List all the potential outcomes of this decision, including the negative ones."

    - "What information would I need to change my mind about this decision?"

    - "If I were advising someone else on this decision, what would I tell them?"

- **For Problem-Solving**

    - "Brainstorm alternative solutions to this problem, even if they seem unlikely."

    - "What are the potential flaws in my proposed solution?"

    - "How could I test my assumptions about this problem?"

Using Prompts with AI

When using these prompts with AI tools, be specific and provide context. For example, instead of asking "What are the arguments against my position?", try "I believe that electric cars are the future of transportation. What are the strongest arguments against this position, considering factors like cost, infrastructure, and environmental impact?"

By actively using these tools and prompts, you can become more aware of your own biases and make more informed decisions.

Remember: These tools aren't here to change your mind – they're here to help you make sure you're right for the right reasons. Use them as your personal reality-check team.

Think of it like having a gym buddy for your brain. Just as we need someone to spot us when lifting weights, these AI tools spot us when we're lifting ideas, making sure we don't hurt ourselves with sloppy thinking.

Ready to put your beliefs to the test? Pick one thing you're absolutely sure about, and let these tools challenge you. You might be surprised at what you discover.

Breaking Free from the Group Mind: Your AI Guide to Better Team Decisions

We've all been there: sitting in a meeting, knowing something doesn't feel right, but staying quiet because everyone else seems on board. Well, it's time to shake things up. Let me show you how AI can be your secret weapon against groupthink.

Your Digital Toolbox for Bolder Thinking

AI tools can be highly effective in helping to overcome groupthink by providing unbiased suggestions, offering data-driven insights, and encouraging diverse viewpoints. Here's a list of AI tools and prompts that can support this goal. Be aware that some of the tools listed (such as SurveyMonkey) have a significant cost and learning curve. These may be best for the enterprise rather than the individual. Experiment with free tools first!

AI Tools to Overcome Groupthink

Tools to depersonalize the debate

- **Text Analysis and Sentiment Tools**
    - ○ **GPT-4/ChatGPT**: Helps rewrite arguments to remove emotionally charged language and make them more neutral.
    - ○ **IBM Watson Tone Analyzer**: Analyzes the tone of a discussion and suggests ways to neutralize overly emotional or personal tones.
- **Sentiment Analysis Tools** (e.g., **MonkeyLearn, Lexalytics**): These tools can analyze text from group discussions or meeting transcripts to detect signs of groupthink, such as suppression of dissenting opinions or excessive agreement.

- **Language Refinement and Rewriting**
  - **QuillBot**: Paraphrases and rewrites text in a neutral tone.
  - **Grammarly**: Identifies tone issues and suggests adjustments to make arguments more neutral.
  - **Wordtune**: Helps reframe ideas and adjust the tone to fit a neutral style.
- **Debate Moderation Platforms**
  - **Kialo**: A platform for structured debates that focuses on clarity and idea exploration rather than personal opinions.
  - **DebateGraph**: Visualizes arguments and counterarguments in a depersonalized, structured format.
  - **Parlay Ideas**: Encourages constructive classroom debates and ensures discussions remain focused on ideas.
- **AI-Powered Mediation Tools**
  - **ODR Platforms (Online Dispute Resolution)**: Tools like **Modria** are used in mediation to maintain objectivity and focus discussions on resolutions rather than personal differences.
  - **Replika**: Although primarily a chatbot for personal conversations, it can simulate arguments to practice depersonalized debate techniques.

- **Argument Mapping and Visualization Tools**
  - ○ **Rationale:** Helps map out logical structures of arguments, clarifying key points and evidence.
  - ○ **MindMup:** Creates mind maps to organize ideas without attaching personal biases.
  - ○ **Arguman:** A web-based tool for mapping arguments, focusing on premises and conclusions.

Other tools

- **Anonymous Feedback Platforms:** Tools like **SurveyMonkey, Google Forms** (with anonymous settings), or dedicated feedback platforms allow individuals to share their opinions and concerns without fear of reprisal or social pressure.

- **Collaborative Document Editing Software:** Platforms like **Google Docs** or **Microsoft Teams** allow for simultaneous editing and commenting, which can facilitate open discussion and brainstorming. However, care must be taken to ensure all voices are heard and dominant personalities don't overshadow others.

- **AI-Powered Decision-Making Platforms: Peltarion** or **DataRobot:** These AI platforms use machine learning models to evaluate the risks and rewards of different decisions.

- **Virtual Meeting Platforms with Breakout Rooms:** Tools like **Zoom** or **Microsoft Teams** allow for smaller group discussions, which can encourage more open communication and reduce the pressure to conform to the majority opinion.

Prompts

These prompts can be used in group settings or as individual reflection exercises before or after group discussions:

General Prompts

- "What are the potential risks and downsides of this decision/plan?"

- "What are we *not* considering?"

- "If we were trying to sabotage this plan, how would we do it?" (This can be a fun way to uncover potential weaknesses.)

- "What data or evidence would change our minds about this?"

- "Are we prematurely converging on a solution? Have we explored enough alternatives?"

- "If someone outside our group were evaluating this, what criticisms might they have?"

- "Let's play devil's advocate. What are the strongest counterarguments to this plan?"

Specific Prompts for Meetings

- "Before we make a decision, let's each take a few minutes to write down our individual thoughts and concerns."

- "Let's have a designated 'devil's advocate' for this discussion." (Rotate this role.)

- "Let's use a round-robin format to ensure everyone has a chance to speak."

- "Let's focus on identifying assumptions we are making and testing their validity."

- "Are there any dissenting opinions? We want to hear them." (Create a safe space for dissent.)

- "What information are we missing that could help us make a better decision?"

Prompts for Post-Meeting Reflection

- "Did everyone feel comfortable expressing their opinions during the meeting?"

- "Were there any points of disagreement that were not fully explored?"

- "Did the discussion feel rushed or pressured in any way?"

- "Did any one person or group dominate the discussion?"

Using Prompts with AI

You can use these prompts with AI tools to

- Analyze meeting transcripts for signs of groupthink (e.g., lack of dissenting opinions, use of pressure tactics)

- Generate alternative solutions or counterarguments to proposed plans

- Simulate different scenarios and outcomes to test the robustness of decisions

By actively using these tools and prompts, teams and groups can create a more open and inclusive environment, leading to better decisions and outcomes. It's about fostering a culture of psychological safety where dissenting opinions are valued and explored.

Remember: The goal isn't to create conflict – it's to make better decisions by hearing all voices, even the artificial ones. Think of AI as your impartial meeting moderator, making sure no good idea gets lost in the shuffle of social dynamics.

Ready to shake up your next team meeting? Start small: try one AI tool to gather anonymous feedback. You might be surprised at what people really think when they feel safe sharing.

The best teams aren't afraid of different opinions – they thrive on them. Let AI help you build that kind of team.

Winning Over Your Fear of Loss: Your AI Guide to Bolder Decisions

Ever pass up a great opportunity because you were too scared of what might go wrong? You're not alone. Our brains are wired to fear losses about twice as much as we value gains. But here's the good news: artificial intelligence can help you overcome this mental speed bump. Be aware that some of the tools listed (such as **Crystal Ball**) have a significant cost and learning curve. These may be best for the enterprise rather than the individual. Experiment with free tools first!

Tools

- *AI-Powered Decision-Making Platforms:* **Peltarion, DataRobot**: These AI platforms use machine learning models to evaluate the risks and rewards of different decisions.
- *Risk Assessment AI Tools:* **Crystal Ball** (by Oracle), **@Risk** (by Palisade): These AI tools use simulations and Monte Carlo methods to assess the risk and uncertainty of a decision.
- *Spreadsheet Software:* **Excel, Google Sheets**: You can use spreadsheets to create decision matrices, calculate expected values, and visualize potential outcomes.
- *AI-Powered Financial Planning Tools:* **FinChat, Truewind**: These AI tools analyze behavioral data to detect biases like loss aversion in decision-making.

- *Scenario Planning AI*
  AnyLogic: These tools use AI to create simulations of different scenarios, helping you visualize how choices will play out in both positive and negative circumstances.

- *AI-Powered Sentiment and Emotional Analysis*
  MonkeyLearn: These AI tools can analyze your written or spoken words for emotional sentiment, helping you recognize if loss aversion is influencing your language or decisions.

- *AI-Powered Journaling and Reflection Tools*
  Daylio, **Reflectly**: AI-driven journaling apps can help track your mood and decisions over time, providing insights into how loss aversion might be affecting your decision-making.

Prompts

These prompts can be used for self-reflection or when discussing decisions with others:

General Prompts

- "What are the potential gains of taking this risk?" (Focus on the positive.)
- "If this decision results in a loss, what are the worst-case and best-case scenarios?" (Frame the loss.)
- "How does the potential loss compare to the potential gain in percentage terms?" (Quantify the difference.)

- "If I had already acquired this [item/outcome], would I be willing to sell it for this price/give it up for this outcome?" (Endowment effect reversal.)
- "Am I focusing too much on avoiding a potential loss and neglecting potential gains?"
- "What is the long-term impact of this decision, rather than focusing on the immediate potential loss?"
- "If I were advising a friend in this situation, what would I tell them?" (Distance yourself from the situation.)
- "What is the opportunity cost of *not* taking this risk?" (Focus on the potential loss of inaction.)

Specific Prompts for Investment Decisions

- "What is my overall investment strategy, and how does this decision fit into it?" (Long-term perspective.)
- "What is the historical volatility of this investment, and am I comfortable with that level of risk?" (Objective data.)
- "Am I making this decision based on fear of missing out (FOMO) or a well-reasoned analysis?" (Emotional check.)

Specific Prompts for Business Decisions

- "What is the potential return on investment (ROI) of this project?" (Quantifiable gain.)
- "What are the alternative uses for the resources I would invest in this project?" (Opportunity cost.)

- "What is the probability of success for this project, based on market research and other data?" (Objective assessment.)

Using Prompts with AI

You can use these prompts with AI to

- Generate alternative framings of potential outcomes. For example, instead of "You could lose $100," the AI could suggest "You have an 80% chance of gaining $200 and a 20% chance of losing $100."
- Calculate expected values and other relevant metrics to help you objectively compare potential gains and losses.
- Simulate different scenarios and outcomes to help you understand the range of possibilities.

By consciously using these tools and prompts, you can become more aware of how loss aversion is influencing your decisions and make more rational choices. The key is to shift your focus from solely avoiding losses to also considering the potential for gains and the long-term consequences of your choices.

Remember: The goal isn't to become reckless. It's about finding that sweet spot between playing it too safe and taking foolish risks. Think of AI as your emotional spotter – helping you lift heavier weights while keeping you from getting hurt.

Ready to face those fears? Start small: Pick one decision you've been avoiding out of fear. Run it through one of these AI tools. You might find that what looks like a scary leap is actually just a manageable step forward.

After all, sometimes the biggest risk is taking no risk at all.

Breaking Free of Narrow Options: Your AI-Powered Guide

Ever feel stuck seeing just one way forward? You're not alone. But here's the good news: artificial intelligence can be your personal thinking coach, helping you spot options you never knew existed. Let me show you how to use AI as your creative partner in decision-making.

Note: as for other types of error some of the tools can be costly and have a steep learning curve. Some such as Miro have a free trial period. Dip your toe in the water and see what works for you.

Tools

- *Mind Mapping Software*
  XMind, MindManager, Miro: These tools help visually organize thoughts and explore related concepts, encouraging broader thinking.

- *Brainstorming Tools*
  Miro, Mural, Stormboard: These collaborative platforms facilitate group brainstorming and idea generation, promoting diverse perspectives.
- *Scenario Planning Software*
  - **Worst Case Scenario**: An AI tool that predicts possible worst-case outcomes for various scenarios.
  - **Plan Quest**: An AI-based tool designed to facilitate project planning and execution. It helps you create detailed plans and manage projects efficiently.
- *Decision Matrix Software*
  MyMap.ai, **Asana** decision matrix template. These tools can help you make more informed and objective decisions by organizing and evaluating multiple options against various criteria.

- *AI-powered Idea Generation Tools*
  - **HyperWrite** Idea Generator: A tool that uses advanced AI technology to generate unique and innovative ideas.
  - **Voilà** Idea Generator: A free online tool that helps overcome creative blocks by generating fresh ideas for projects, stories, or brainstorming sessions.
  - **Musely** Idea Generator: This tool transforms creative blocks into endless possibilities by

generating ideas for blog topics, startup concepts, and creative projects

- *Marketing*
    - ○ **Jasper AI**: Offers tailored suggestions for marketing, writing, and business ideas.
    - ○ **Writesonic**: Provides creative suggestions for content creation, marketing, and storytelling.
    - ○ **Rytr**: Helps create content ideas and drafts for various purposes, including marketing and storytelling.

- *Business planning*
    - ○ **IdeaBuddy**: Helps generate, refine, and structure business ideas into actionable plans.
    - ○ **Notion AI**: Suggests creative and strategic ideas for organizing workflows, projects, or goals.

- *Other*
    - ○ **You.com** provides a more general platform that can be customized.
    - ○ **DebateGraph**: Structured argumentation and counterargument visualization.
    - ○ **Arguman:** Argument mapping for personal or team decisions.

Prompts

These prompts can be used for individual reflection or in group discussions:

General Prompts

- "What is the actual problem we're trying to solve?" (Challenge the initial framing.)
- "What are the broader goals or objectives we're trying to achieve?" (Focus on the bigger picture.)
- "What are other ways to define this problem?" (Reframe the issue.)
- "What are some unconventional or 'out-of-the-box' solutions?" (Encourage creative thinking.)
- "If we had unlimited resources, what would we do?" (Remove constraints.)
- "What would someone from a different background or perspective suggest?" (Seek diverse viewpoints.)
- "What are the long-term consequences of each potential solution?" (Consider the future impact.)
- "What other related problems or opportunities could this address?" (Expand the scope.)

Specific Prompts for Problem Definition

- "What are the root causes of this problem, not just the symptoms?" (Dig deeper.)
- "What are the stakeholders' needs and perspectives?" (Consider different viewpoints.)
- "What are the constraints and limitations we're operating under?" (Acknowledge realities.)

- "What are the positive aspects of the current situation that we want to preserve?" (Balance change with stability.)

Specific Prompts for Solution Generation
- "What are some solutions that have worked in other contexts or industries?" (Borrow ideas.)
- "What are the trade-offs associated with each potential solution?" (Consider the costs and benefits.)
- "How could we combine different solutions to create a more comprehensive approach?" (Integrate ideas.)
- "What are the potential unintended consequences of each solution?" (Anticipate problems.)

Using Prompts with AI

You can use these prompts with AI to

- Generate alternative problem definitions based on different perspectives or contexts.
- Brainstorm a wider range of solutions based on the given problem definition.
- Analyze the potential consequences of different solutions, considering various factors and scenarios.
- Identify analogous problems in other domains and suggest potential solutions based on those analogies.

By actively using these tools and prompts, you can break free from narrow framing, explore a broader range of possibilities, and make more informed and effective decisions. The key is to consciously challenge your initial assumptions and seek out diverse perspectives.

Remember, AI isn't here to make decisions for you – it's your thought-expanding teammate. Use it to explore what-ifs, challenge your assumptions, and discover possibilities hiding in your blind spots.

The Next Step

Start small. Take one decision you're wrestling with right now. Pick one AI tool and ask it three questions you wouldn't normally ask yourself. You might be surprised at what you discover.

Ready to break free from mental tunnels and see the whole landscape of possibilities? Your AI thinking partners are waiting to help you explore.

General Principles

Given the rapid evolution of AI technology, certain tools discussed may have undergone significant changes since publication. Additionally, some sophisticated AI solutions may be better suited for corporate environments, considering their subscription costs and complexity of implementation. This discussion primarily addresses queries using AI platforms that offer free access tiers.

You'll likely discover that particular AI tools and questioning techniques resonate more effectively with your needs and thinking style. The optimal approach involves exploring various options while avoiding the temptation to utilize every available tool. Instead of attempting comprehensive coverage, focus on identifying the methods that yield the most valuable insights for your specific situation.

Remember that attempting to synthesize multiple AI outputs simultaneously can create unnecessary complexity. Our cognitive limitations make it challenging to effectively process and integrate numerous AI responses at once. Rather than trying to harmonize multiple AI inputs, concentrate on mastering the tools and approaches that consistently deliver helpful results for your particular needs.

Identifying and flagging biases

Use AI as a systematic thinking partner that can help identify potential biases in your reasoning. Have it analyze your arguments and conclusions for common fallacies like confirmation bias, availability bias, or emotional reasoning.

Finding alternative perspectives

As we have seen in the confirmation bias sections, we are prone to searching for information that confirms our existing biases. Ask AI to do the opposite.

Use AI as the Devil's Advocate: "[my hypothesis]. Give me arguments that go against this".

Use AI to help start a Pre-Mortem[17]: "[describe your plans]. We are a year from now. What went wrong?"

Use AI to generate more options if you are dealing with one (should I do this or not) or two (should I do this or that).

Dealing with complex issues

Break down complex problems into smaller components with AI's help. This can counter the tendency to oversimplify or jump to conclusions. AI can guide you through methodical analysis of each element.

How likely is this explanation?

Use AI to check your probabilistic reasoning. Have it help you think through base rates, conditional probabilities, and statistical concepts that humans often struggle with intuitively.

---

[1] Klein, G., 2003. Intuition at work. Doubleday, New York, NY.

Iterative refinement

Iteratively test and refine prompts to improve accuracy and relevance, reducing errors over time.

This is important as you will realize that AI is good at some things and not at others. What is easy for you may not be so for AI and vice versa. Explore!

Perspective-Taking Through AI Role-Playing

Engage with AI by assigning specific roles and scenarios. Begin with a clear prompt like: "You are a high school teacher dealing with increasing student phone use during class. What approaches could you consider?" or "As a police officer responding to a neighborhood dispute, what options do you have?"

You can also tap into historical wisdom by asking AI to channel notable figures: "How might Marie Curie approach this research setback?" or "What strategies would Steve Jobs use to reimagine this product?" This technique helps you view challenges through the lens of innovative problem-solvers.

For organizational challenges, prompt AI to draw from business success stories: "What creative solutions have successful companies implemented when facing similar customer service challenges?" This approach helps uncover proven strategies while encouraging novel applications.

These role-playing techniques serve two primary purposes. First, they help overcome processing errors by presenting situations from fresh angles. Second, they provide valuable insights for navigating complex social dynamics and interpersonal situations.

Additionally, you can use role-specific prompts to tailor content for different audiences. For instance, request: "As a fifth-grade science teacher, explain climate change to your students" or "As a corporate communications director, craft a press release about this technology update for industry professionals." Remember to include relevant context with each query, such as audience background, current circumstances, or specific constraints.

By providing clear context and specific roles, you help AI generate more focused, practical, and audience-appropriate responses. This approach transforms abstract problems into concrete scenarios with actionable solutions.

Leveraging Multiple AI Tools

Different AI systems often excel in distinct areas. You might discover that one particular AI demonstrates superior capability in generating comprehensive lists, while another shows remarkable skill in crafting narrative text. Take time to identify which tools serve you best in specific scenarios.

Consider expanding your problem-solving toolkit by employing multiple AI resources. However, this approach comes with important considerations. The first is time investment - consulting multiple systems requires additional effort. The second relates to our cognitive limitations. While you might be tempted to combine outputs from different AIs, merging multiple lists or complex suggestions proves challenging due to the constraints of human working memory. Our prefrontal cortex can only actively process a limited number of items simultaneously.

Instead of attempting to synthesize multiple outputs, consider a strategic approach. Direct different types of queries to the AI tools that handle them most effectively. For example, use one AI for brainstorming sessions and another for detailed analysis or writing tasks.

You can also implement a sequential approach. Take the output generated by one AI system and feed it into another. This method proves particularly valuable when you want to refine initial ideas, obtain critical feedback, or adapt content for specific audiences. For instance, use one AI to generate initial content, then ask another to evaluate it or modify it for a particular reader demographic.

This layered approach allows you to harness the unique strengths of different AI tools while working within the natural limitations of human cognitive processing.

When should we use AI?

This is a difficult question to answer. Here is a summary of Appropriate reliance on GenAI: research synthesis[18]. This summary was AI-generated and is indented and in italic to reflect this.

The report examines how users can develop appropriate reliance on Generative AI (GenAI) systems - accepting correct outputs while rejecting incorrect ones. This is particularly challenging with GenAI due to several unique characteristics:

Key Challenges:

- *Non-deterministic outputs (same input can produce different results)*

- *Systems can make mistakes when questioned about their accuracy*

- *Systems can be influenced by user input style and perceived user characteristics*

- *The speed and fluency of output can mask errors*

- *High verification costs for users*

---

[18] Passi, S., Dhanorkar, S., & Vorvoreanu, D., 2024. Appropriate reliance on GenAI: Research synthesis - https://www.microsoft.com/en-us/research/uploads/prod/2024/03/GenAI_AppropriateReliance_Published2024-3-21.pdf

The report identifies three main factors affecting overreliance:

1. *User expertise levels*

2. *Type of interaction (single vs. multi-turn conversations)*

3. *Task type (e.g., coding vs. creative writing)*

Effective mitigation strategies include:

- *Explanations:*
    - *Verification-focused explanations helping users assess output correctness*
    - *Background explanations providing external context*
    - *Note: Explanations can backfire by increasing overreliance if poorly implemented*
- *Uncertainty Expressions:*
    - *Visual indicators (e.g., highlighting uncertain content)*
    - *Linguistic expressions (e.g., "I'm not sure...")*
    - *Particularly effective when using first-person expressions*
    - *Caution: Model confidence doesn't always correlate with accuracy*
- *Cognitive Forcing Functions:*
    - *Self-critiques by the AI system*
    - *Questions promoting critical thinking*
    - *Can help users shift from automatic to more deliberative thinking*

- Risk: May lead to under-reliance or increased cognitive burden

Design Recommendations:

1. *Be transparent about system capabilities and limitations*

2. *Provide relevant explanations to help users verify outputs*

3. *Clearly convey model uncertainty*

4. *Use cognitive forcing functions appropriately for the user and context*

The report emphasizes that implementation of these strategies should be carefully tested in context, as inappropriate application can backfire and reduce effectiveness. Success requires balancing the need for accuracy verification with user experience and cognitive load.

This research is particularly relevant for developing systems that help counter thinking errors, as it provides frameworks for encouraging more deliberative thought processes and critical evaluation of AI outputs.

Additional Considerations for AI Implementation

Before diving into AI-assisted analysis, consider the investment of time and effort required. Evaluate whether your situation merits such detailed examination. Not every decision requires comprehensive AI consultation. Assess the potential impact and complexity of your choice to determine if AI analysis would provide meaningful value.

Keep in mind that the AI landscape evolves rapidly. A tool that previously fell short of your needs may now offer enhanced capabilities. Regular reassessment of available AI resources ensures you're leveraging the most current and effective solutions.

When employing AI tools, focus on these key applications:

First, use AI to examine situations from multiple perspectives, helping you break free from fixed viewpoints and consider alternative approaches you might not have otherwise contemplated.

Second, leverage AI to gain insights unbounded by your immediate social or professional circle, providing fresh perspectives that might challenge or complement your group's collective thinking.

Third, utilize AI to help identify emotional factors that might influence your decision-making, serving as an objective observer of potential emotional biases.

Finally, employ AI to expand or contract your option set as needed. When facing limited choices, use it to generate additional possibilities. Conversely, when overwhelmed by too many options, use it to help categorize and prioritize your choices into manageable alternatives.

Understanding AI Inaccuracies

Like human memory, AI can generate inaccurate information, sometimes creating fictional narratives or data. However, you can employ several proven strategies to identify potential inaccuracies in AI responses.

When evaluating AI output, begin by comparing specific claims against reputable sources. If information appears in an AI response but cannot be verified through trusted references, approach it with caution. This verification process becomes particularly important for factual claims, historical events, or statistical data.

Pay attention to logical coherence within AI responses. Fabricated information often reveals itself through internal contradictions. For instance, you might notice chronological inconsistencies or details that shift as the response progresses.

Watch for hyper-specific information, particularly regarding obscure topics. When AI provides unusually precise details about lesser-known events or subjects, these details warrant additional verification. This is especially true for exact dates, statistics, or numerical data about historical events that lack extensive documentation.

When AI references specific publications or research, verify these sources independently. Remember that AI's ability to cite accurately may be limited by its training data and lack of real-time access to academic databases.

Approach unique claims with skepticism. If information provided by AI seems unavailable through other channels or contradicts established knowledge in a field, subject it to thorough verification.

Consider the AI's level of certainty in its responses. When discussing specialized or niche topics where reliable information is scarce, appropriate expressions of uncertainty indicate more reliable AI behavior than unwavering confidence.

Request additional examples or illustrations of key points. Apply the same verification principles to these examples to assess their accuracy.

A particularly effective approach involves using multiple AI systems in a structured manner. Research indicates that employing three different AI platforms systematically can significantly reduce the occurrence of inaccurate information[19].

Additional tools

Emotional state tracking applications such as **Moodfit** or **Daylio** can serve as valuable early warning systems, alerting you when your mental or emotional condition might compromise your decision-making abilities. These tools complement the age-old wisdom of allowing time for

---

[19] Gosmar, D., & Dahl, D.D., 2025. Hallucination mitigation using agentic ai natural language-based frameworks. Submitted/Preprint, https://arxiv.org/pdf/2501.13946

reflection before making significant choices. Just as our ancestors understood the value of 'sleeping on it," modern technology can help us recognize when we might benefit from postponing important decisions.

However, it's crucial to understand the proper role of these AI-powered assistants in your decision-making process. While they offer valuable support in developing better thinking habits and identifying potential cognitive biases, they should not be viewed as substitutes for professional guidance when such expertise is needed. Consider these tools as cognitive training aids - similar to how training wheels help develop balance on a bicycle, these applications help you recognize and correct thinking patterns that might otherwise lead to judgment errors.

These tools work best as part of a broader approach to decision-making, complementing rather than replacing human judgment and professional expertise when needed.

Applications

Our brains love to play tricks on us. Sometimes we catastrophize, seeing disasters around every corner. Other times we filter out all the good stuff and focus on that one negative comment. But here's some good news: artificial intelligence can be your personal reality checker, helping you catch these mental slip-ups before they catch you.

Let's break down how AI can help tame those pesky thought patterns in various areas of our lives. By now you should be able to generate useful prompts. Here are tools for different situations:

Relationships

- Mind Reading
    - *AI Chatbots*: (e.g., **Replika**) Simulate conversations to practice healthy communication skills.
    - *Emotion Analysis*: Tools like **MonkeyLearn** are user-friendly and help differentiate assumptions from facts.
- Catastrophizing
    - *CBT-Based Apps*: Apps like **Woebot** guide users through reframing catastrophic thoughts.

- Predictive Analysis Tools: **Microsoft PowerBI** is mostly designed for companies rather than individuals.

- All-or-Nothing Thinking

  - *Cognitive Journals*: Apps like **Thought Diary** track nuances in situations to promote balanced thinking.

  - *AI Feedback*: Virtual coaches such as **Rocky.ai** offer perspectives to avoid extreme judgments.

- Overgeneralization

  - *Pattern Recognition Tools*: **ThinkBetter Academy**: Offers techniques for enhancing pattern recognition skills, such as focusing on similarities and differences, utilizing mnemonics, and solving puzzles AI identifies recurring thinking errors in journal entries. A pitiful website but interesting information.

- Personalization

  - *AI Sentiment Tracking*: **Chat Recap AI**: Analyzes conversations to uncover emotions and relational dynamics, offering actionable insights to improve communication.

Work or School

- Perfectionism

    - *AI Goal Setting Tools*: Tools like **Coach.me** encourage setting realistic goals.

    - *Error Analysis Tools*: Apps provide constructive feedback without overemphasizing perfection.

        - **Replika:** A virtual companion that helps users practice expressing emotions and refining communication skills, focusing on personal growth rather than perfection.

        - **Texta AI Writer:** Offers instant feedback on writing to enhance clarity and coherence without focusing solely on perfection.

- Discounting the Positive

    - *Gratitude Journals*: AI-driven prompts (e.g., in **Happify**) focus on acknowledging achievements.

    - *AI-Generated Highlights:*

        - **ClickUp:** A multifaceted project management tool that offers extensive features for tracking tasks, deadlines, and progress, with AI-driven insights to highlight positive developments.

- **Wrike**: Known for its robust scheduling and reporting features, Wrike uses AI to provide comprehensive project insights and highlight positive progress.

- **Zoho Projects**: This cloud-based tool uses AI to automate task assignments, predict project deadlines, and optimize workflows, making it easier to identify and summarize positive progress.

- Imposter Syndrome

(When even successful people feel like they're "faking it" and doubt their achievements despite evidence of their abilities)

- *AI Mentorship Platforms*

  - **GrowthMentor**: Connects you with experienced mentors who have overcome imposter syndrome themselves, offering personalized advice and support.

  - **YesChat Imposter Syndrome**: An AI-powered mentorship tool that provides tailored advice and strategies to help you overcome self-doubt and build confidence.

- Fortune Telling

  o *Scenario Simulation:* Simulation Software tools like **AnyLogic** and **Simul8** allow users to create detailed simulations of real-world processes and scenarios, providing insights into how different factors might influence outcomes.

Health and Wellness

- Overgeneralization

  o *Habit Trackers*: AI apps like **Trackhabit** and **Habitbull** show overall progress rather than isolated failures.

- Magnification

(A type of thinking that involves exaggerating the negative aspects of oneself, others, or a situation while downplaying the positive ones).

  o *Behavioral Analytics*: Data-driven insights contextualize minor setbacks within broader success. **Mindfulmate** offers AI-powered therapy to help users recognize and manage cognitive distortions like magnification and minimization.

- Should Statements

  - *AI Coaching*: Apps like **BetterUp** provide personalized guidance to replace "should" with practical advice.

- Black-and-White Thinking

  - **Mindfulmate**: Offers AI-powered therapy to help users recognize and manage cognitive distortions like black-and-white thinking.

- Emotional Reasoning

  - *Emotion Detection*: Turn your computer camera on yourself, talk about what you are experiencing, and use one of several video emotion analysis tools:
    - **Affectiva**: Uses AI to analyze facial expressions and vocal intonations to detect emotions in real-time, commonly used in market research and customer experience analysis.
    - **Emotient**: Focuses on analyzing facial expressions to understand emotions, often used in advertising and media to gauge audience reactions.
    - **Realeyes**: An AI-powered tool that measures emotional responses to video content by analyzing facial expressions and eye movements.

Social Situations

- **Mind Reading:** *Conversation Simulators*: Practice realistic social scenarios with AI bots such as **YesChat**.

- **Catastrophizing:** *AI Cognitive Restructuring*: Tools like **Youper** guide users through calming, evidence-based responses.

- **Spotlight Effect**[20]: The psychological phenomenon by which people tend to believe they are being noticed more than they really are. Anxiety management apps such as **Headspace** can help.

- **Comparing**

  Self-Progress Tracking: Apps highlight personal growth instead of external comparisons.

  - **ClickUp**: A comprehensive goal setting and management tool with customizable task management capabilities and pre-built templates for quick goal setting.

---

[20] Gilovich, T.; Medvec, V. H.; Savitsky, K., 2000. The Spotlight Effect in social judgment: An egocentric bias in estimates of the salience of one's own actions and appearance. Journal of Personality and Social Psychology. 78 (2): 211–222. doi:10.1037//0022-3514.78.2.211.

- Strides: Offers habit trackers and goal-setting features to help you establish and maintain productive routines.

- Way of Life: Focuses on cultivating good habits and consciously removing undesirable patterns, with a built-in diary for identifying habit triggers.

- Self-Blame

  Here are some AI tools designed to help counter self-blame and promote self-compassion:

  - Woebot: An AI-powered chatbot that uses cognitive-behavioral therapy (CBT) techniques to help users manage negative thoughts and self-blame.

  - Replika: A virtual companion that offers supportive conversations and helps users practice self-compassion and positive self-talk.

  - Mindfulmate: Offers AI-powered therapy to help users recognize and manage cognitive distortions, including self-blame Decision-Making

- Analysis Paralysis

  - *Decision Support Tools*: AI tools like **Otter.ai** organize options and highlight priorities.

- **Confirmation Bias**

  AI-powered debate platforms such as **Kialo** and **Polis** can expose alternative views. See section in previous chapter for more options.

- **Overconfidence**

  Here are some AI tools designed to help counter overconfidence and promote more balanced decision-making:

  - **Mindfulmate:** Offers AI-powered therapy to help users recognize and manage cognitive distortions, including overconfidence.

  - **Replika:** A virtual companion that offers supportive conversations and helps users practice self-awareness and balanced thinking.

  - **Woebot:** Uses cognitive-behavioral therapy (CBT) techniques to help users manage negative thoughts and overconfidence.

  - Also consider that the main source of overconfidence is how we process information, with emotions also playing a role[21]. See the sections about confirmation

---

[21] Golard, A, 2021. A Field Guide to Thinking Errors, pp167-168.

bias and loss aversion for more tools as well as prompts.

- Loss Aversion

Here are some AI tools designed to help counter loss aversion and promote more balanced decision-making:

  o The tools listed for overconfidence apply here.

  o Also see loss aversion section in previous chapters.

- Hindsight Bias

  o The tools used for overconfidence apply here as well.

  o Also consider that the source of this error is how we process information[22], so the section on confirmation bias can also provide insights.

General Emotional Challenges

- Negative Filtering

  o **Mood Trackers**: Apps like **Moodfit** highlight positive patterns in daily life.

---

[22] Golard, A, 2021. A Field Guide to Thinking Errors, pp113-114.

- Over-Attachment to Thoughts

  - **Mindfulness Apps**: AI-guided meditation tools (e.g., **Headspace**) teach detachment from negative thoughts.

- Rumination

  - Time tracking apps such as **Rescuetime** and **TogglTrack** can recommend activities to interrupt rumination patterns.

- Low Frustration Tolerance

Low frustration tolerance (LFT) is a concept utilized to describe the inability to tolerate unpleasant feelings or stressful situations. Here are some AI tools to counter it:

  - **YesChat** Frustration Tolerance Trainer: Engages users in challenging conversations and scenarios to train and enhance their ability to tolerate frustration.

  - **Mindfulmate**: Provides AI-powered therapy to help users recognize and manage cognitive distortions, including building resilience against frustration.

- Global Labeling

  - *AI Writing Feedback*: Identify and suggest replacements for overgeneralizing language in journaling or notes. Here are some AI tools designed to help counter overgeneralizing language and promote more balanced thinking:

- **Mindfulmate**: Provides AI-powered therapy to help users recognize and manage cognitive distortions, including overgeneralizing language.

- **Replika**: A virtual companion that offers supportive conversations and helps users practice self-awareness and balanced thinking.

Conclusions

Identifying your personal cognitive biases requires careful self-reflection. Start by examining how you reach conclusions. When an answer seems to come instantly, pause and consider alternative perspectives. Within group settings, assess whether you feel pressure to conform or if certain voices are being suppressed. Consider whether fear of isolation influences your choices.

A powerful technique is to verbalize your situation while looking in a mirror. Pay attention to your emotional responses as you do this. Examine your decision-making process: Are you limiting yourself to too few options, or are you overwhelmed by too many? If facing limited choices, challenge yourself to generate alternatives. If drowning in possibilities, look for ways to categorize and simplify them.

Artificial intelligence can be a valuable partner in all these scenarios. Refer to the Let AI Be Your Mental Detective chapter for specific techniques and prompting strategies. The Applications section also provides practical entry points for using AI in this context.

One of AI's greatest strengths is its ability to introduce perspectives from beyond your immediate social circle and personal experience. It particularly excels at helping with emotionally driven cognitive errors, such as loss aversion. Unlike humans, who become more prone to these errors when fatigued, stressed, or distracted[23], AI maintains consistent performance regardless of circumstances.

For decisions you make regularly, consider developing a systematic approach to tracking AI interactions. Create a matrix that records four outcomes: when you accepted AI advice and it proved correct or incorrect, and when you rejected AI advice and it proved correct or incorrect[24]. While this requires dedication and time investment, it can provide valuable insights into your decision-making patterns. Remember that AI capabilities are continuously improving, so your evaluation metrics may need periodic adjustment.

Don't feel constrained by the tools mentioned in this book – they're meant as a starting point. The AI landscape evolves rapidly, with new resources emerging constantly. When seeking specific solutions, ask large language models to suggest relevant tools for your needs.

Consider the suggested prompts as valuable questions for meetings, whether or not you're using them with AI. Incorporate AI as well to enhance your approach. This strategy creates a seamless connection between AI

[23] Shiv, B., & Fedorikhin, A., 1999. Heart and Mind in Conflict: The Interplay of Affect and Cognition in Consumer Decision Making. The Journal of Consumer Research 26, 278-292.
https://www.jstor.org/stable/10.1086/209563
[24] Appropriate reliance on GenAI: Research synthesis -
https://www.microsoft.com/en-us/research/uploads/prod/2024/03/GenAI_AppropriateReliance_Published2024-3-21.pdf

utilization and leadership development. You might view AI as another powerful resource in your leadership toolkit.

Exercise caution: While AI readily generates lists and suggestions, not all will be relevant, and some may be inaccurate. Think of AI as a powerful but imperfect tool. For crucial decisions, employ multiple AI tools and varied prompts, just as you might seek advice from several trusted friends.

Keep in mind that AI is in constant evolution. Optimal prompting strategies may shift as new reasoning models emerge. Maintain an experimental mindset, testing different approaches across various tools. Despite this ongoing change, the fundamental principles outlined in this book provide a solid foundation for leveraging AI to overcome cognitive biases.

Index

Andre Golard received his PhD from New York University and an Executive MBA from the University of Washington. He was an HHMI Fellow in the Center for Neurobiology and Behavior at Columbia. His career spans academia, startups, and big tech companies. Twice a TEDx speaker, he is an independent consultant and coach based in Seattle. This is his second book.

www.ingramcontent.com/pod-product-compliance
Lightning Source LLC
Chambersburg PA
CBHW070346130626
46556CB00007B/3051